Marine Algae Control Secrets

Remedies for the Marine and Reef Aquarium

by

Bob Goemans, Ph.D.

TABLE OF CONTENTS

Marine Algae Control Secrets

Remedies for the Marine and Reef Aquarium

by

Bob Goemans, Ph.D.

Foreword

Marine algae – some aquarists cultivate it, some curse it. In the seventies and early eighties most aquarists considered healthy growths of micro or macroalgae an indicator of good water quality. In some ways that was correct. In other ways, those growths may have become an annoyance. In fact, there's no doubt that abundant growths of unwanted forms of algae have caused some aquarists to leave the hobby in disgust.

Numerous books have dedicated portions of their text to marine algae. Nevertheless, I doubt whether the subject matter of "causes and control" has ever had the focus as it appears in this booklet.

Most aquarists realize it's difficult to find algae on the reefs in the wild. However, there are always some calcareous and turf algae because they are actually important building components of any reef system. Yet, with the reef's surrounding water being nutrient poor and herbivores keeping what there is of algae well mowed, little or none is readily visible. But it's another story in shallow inland waterways and bays where nutrients have a tendency to accumulate. There, algae in various forms and amounts can easily be seen. It's these types of algae that present the most problems for many aquarists.

As mentioned above I can remember when a good growth of any kind of green algae in a marine aquarium was time for smiles. That's now no longer true in most marine aquariums, especially reef aquariums. Only calcareous algae such as Halimeda or Coralline, or possibly a controlled growth of Caulerpa is considered acceptable. Filamentous or slime-like growths are almost always considered problematic and usually become costly in time and money to remedy.

There are various colors of algae in aquariums, e.g., green, brown, and red, and they can be single-cell, multicellular, encrusting, slime-like, or plant-like in appearance. Some perform beneficial functions such as the production of oxygen and removal of various toxic compounds. Others can be harmful to fishes and coral.

To foster conditions in the closed system that compliment only the more beneficial forms of algae is something that needs to be discussed in a thorough and easily understood manner. That's difficult because there are many contributing factors, types of algae, and misconceptions surrounding the subject matter. However, I'm making it the goal of this booklet.

To begin, let's first take a look at the general plant family, then at what's considered preferable and unwanted. Next, the contributing factors are examined including the generally accepted reasons for proliferation, which need enlightenment.

A checklist containing causes and remedies for different problematic growths is then presented. It's tuned towards concisely listing all causes and summarizing general and more specific remedies. By separating this information, repetition of certain aspects will be extremely limited. In the closing text I'll discuss how to construct a system that will remain free of most unwanted algae. In accomplishing this endeavor I have three goals:

1) Whatever the aquarist decides after reading this booklet, I want the information here to be helpful in accomplishing that decision.
2) I want this booklet to help the aquarist discover why these algae growths occur and persist, and how to control or eradicate them.
3) And finally, help aquarists have long-term success with their aquariums.

In this third of the "Secrets" series, the text has been written for a broad audience hoping to appeal to all that read it. Some areas may be slightly technical, but I believe the broader scope will present a more interesting and informative work.

As with any subject matter, the best place to start is in the beginning, so let's begin with a general look at The Plant World.

The Plant World

Most marine plants/algae are either too large or simply undesirable for aquarium use. Actually, the word "plants" is a misnomer when discussing algae in general. Algae differ from terrestrial plants because they lack true leaves, a root system, and a system to transport nutrients. Since the term "plants" continues to be widely used in the aquarium industry, its use will continue here as appropriate.

Marine plants are members of three of the five plant kingdoms and are defined here in more detail. They are related via their symbiotic relationships with organisms in the two remaining kingdoms. The five kingdoms are:

- Kingdom Monera
  Division Schizophyta – Heterotrophic/cyanobacteria.
  Division Cyanophyta – Blue green algae.
- Kingdom Mycota – Fungi.
- Kingdom Protista
  Division Euglenophyta – Euglenoids; freshwater, marine, and mud.
  Division Pyrrophyta – Dinoflagellates; unicellular, marine, brackish, fresh, and sand. Includes *Amyloodinium,* the "rust-like disease" that infects marine fishes.
  Division Chrysophyta – Golden brown algae.
  Division Bacillariophyta – Diatoms; silicate based exoskeletons, unicellular. Marine, freshwater and soil.
  Division Xanthophyta – Yellow green algae, mostly non-marine.
  Division Phaeophyta – Brown algae/kelp; marine, mostly littoral. About 1,500 species.
  Division Rhodophyta – Red algae; mostly marine, about 4,000 species.
  Division Chlorophyta – Green algae; mostly freshwater, a few marine, some terrestrial.

- Kingdom Metaphyta
- Kingdom Metazoa – Multicellular animals.

Terrestrial seed-bearing plants are contained in the Metaphyta Kingdom. They have roots, true leaves, and transport their nutrients through the plant structure. Most plants in this Kingdom are found on land, yet a few are found in intertidal zones, e.g., turtle grass. In the intertidal zone, besides providing a food source and additional dissolved oxygen, they also provide a temporary yet safe environment for newly hatched fishes.

Of the three closely related kingdoms, the Kingdom Monera and Protista are of more interest to the marine aquarist when it comes to algae control. This Kingdom contains those annoying bacteria slimes often referred to as red-slime algae. The Kingdom Protista has five families of interest. They are:

- Brown Algae (Phaeophyta) – From small encrusting species to that of kelp that can reach 100 feet in height.
- Dinoflagellates (Pyrrophyta) – A cause of red tides.
- Golden Brown Algae (Chrysophyta) – Contains two aquarium-related groups of generally unwanted microalgae.
- Green Algae (Chlorophyta) – A more common green algae found in aquariums containing the pigment called chlorophyll.
- Red Algae (Rhodophyta) – Usually a decorative species for aquariums. The deepest growing of all marine algae. Some have been found in the Bahamas at almost 1000 feet deep.

Beneficial Species

There are many forms of algae that serve a useful purpose in one fashion or another. Unfortunately they can't all be maintained in closed systems. Therefore aquarists need to concentrate on the more commonplace/maintainable species, many of which if properly controlled lend themselves to improving the system's environment. However, if not controlled they can lead to throwing in the towel so to speak!

When accomplishing photosynthesis, algae use carbon dioxide and provide oxygen for fishes and invertebrates. In turn, this helps raise system pH and increase system carbonate buffer capacity (alkalinity). They also help to remove unwanted compounds such as ammonium and nitrate from bulk water, thereby acting like a form of chemical filtration. Keep in mind the nitrate removed is broken down by internal enzymes until the ammonia needed for growth is attained. Algae also serve as a valuable food source for some fishes and invertebrates and provide a decorative aspect or shelter for some inhabitants. Some forms of algae provide important sugar-like compounds for sustaining the hosting animal.

There is even a somewhat beneficial side to some forms of cyanobacteria. And you thought it was all bad! Well, you're almost right, but not 100%. Cyanobacteria is an early colonizer of substrates and an indicator the microbial biomass is progressing. How far these mat-like growths should be allowed to progress is discussed further on in this booklet. As for "more" desirable forms of algae, there are some that are of particular interest to aquarists.

Caulerpa

This green, terrestrial plant-like and multi-shaped macroalgae is probably the most popular. It can be found readily available at most marine aquarium stores. There are about seventy different species of Caulerpa and most are prolific growers. Root-like holdfasts grow along their rhizomes and attach themselves to substrate, rock, and even coral animals. The main purpose of these growths is to anchor the plant to a substrate surface, yet some uptake of nutrients is thought possible. However, the majority of nutrient uptake is through the fron/leaf structure.

Caulerpa are coenocytes, i.e., a multinucleate cytoplasm appearing as many interconnected segments. The proper way to break off a section of the plant is to crush the thallus (stem), not make a clean cut such as with a scissors. Clean cuts can lead to internal damage and quickly cause a section or the specimen itself to disintegrate. Beside disintegration from physical abuse, cellophaning (i.e., becoming transparent) can be caused by sexual reproduction. When this occurs, the Caulerpa leaf will first become blotchy. Soon after blotchiness appears, hair-like discharge tubes called "papillae" will form along the edges of the leaves and discharge gametes and some remaining cytoplasm.

There are some thoughts as to why Caulerpa enter the sexual cycle. Some think it is brought about by salinity changes. Some think it may simply be a biological clock. Others think it may be brought about by a lack of nutrients, e.g., carbon dioxide, nitrate, iron or that of excess organic material. No matter what the cause, when Caulerpa begins to cellophane, those leaves should be removed before disintegrating altogether. If allowed to disintegrate, the plant's nutrients will be released into the aquarium, reducing overall bulk water quality. Excellent growths of Caulerpa can usually be maintained by simply harvesting the oldest one-third of growth as necessary.

Caulerpa, as most algae, does better under regular daylight fluorescent lamps than under more expensive blue spectrum lamps such as 6,500 or 10,000 Kelvin lamps. The reason is that algae prefer longer wavelength light, such as the red band wavelength. Higher Kelvin rated lamps have more blue-green spectrum, which algae do not find as useful as red band. Caulerpa also prefers what could be termed average water conditions, i.e., where nutrients such as ammonia/ammonium, nitrates, and phosphate are readily available.

In fact, I could not get macroalgae to grow in a plenum-equipped system where nitrate was constantly below 3 mg/l nitrate-nitrogen (NO_3-N). Every time I tried a new specimen it would cellophane and die. As an experiment I carefully split a specimen, *C. prolifera*, and added half to another aquarium. This other aquarium had a sandbed directly on its bottom and its temperature and salinity were the same as the plenum-equipped aquarium. The specimen in the aquarium with the sandbed directly on its bottom grew well. The major difference between the two aquariums was the one with sand directly on its bottom had a nitrate level of 8 mg/l. As I brought down its nitrate level, Caulerpa growth slowed. Under 3 mg/l, growth stopped and the specimen began to cellophane.

Some aquarists refer to Caulerpa as a pest because it can block light or its holdfasts annoy corals. Preventing this type of stress is a simple matter of guiding new growth away from the coral animal. It is also known that Caulerpa release a toxin called Caulerpicin that can inhibit coral growth and possibly kill fishes. Caulerpicin and their organic leachates that yellow bulk water can be removed with adequate protein skimming and/or proper use of activated carbon.

Some Caulerpa species are a favorite of herbivore fishes such as Tangs. Notice I said "some" as I've found only two species that are usually eaten with relish. (No, not that kind of relish!) The following four are of more interest:

C. mexicana – Its bright green feather-like growth is the best of all the Caulerpa species for aquarium use. Leaf/fron/blade length varies, yet is usually 6 - 8 inches (15 - 20 cm). The oppositely arranged, segmented points face upward on a basically flat-shaped leaf. The leaves are attached to and spread by rhizomes/runner, which in turn are held in place by holdfasts/rootlets/rhizoid. This form of Caulerpa releases fewer toxins than other types of Caulerpa, and fishes such as Tangs find it quite tasty. In fact, this species and *C. prolifera* are the only two Caulerpa species usually eaten by most herbivorous fishes. Usually referred to as "fern" macroalgae.

C. prolifera – Grows similar to *C. mexicana,* yet has a flat non-segmented tape-like and vertical, slightly curled leaf. The leaf length varies, usually 4 - 8 inches (10 - 20 cm). Sometimes a second leaf will start growing from the main leaf. Most herbivores will eat it. However it's prone to sexual reproduction disintegration if not diligently pruned.

C. racemosa – This light green vine-like Caulerpa has bunches of miniature grape-like beads intertwined on short and low growing branchlets, and forms spreading carpets. The "grapes/spheres" range in size from 1 - 3 mm. Most fishes do not seem to like its taste, yet I have seen Tangs eating it occasionally. Usually referred to as "Grape" macroalgae. This particular species and the one below are highly recommended for lighted refugia where rapid plant growth is beneficial.

C. sertularioides – Small upright blades, about 1 – 2 inches (2 – 5 cm) containing saw-like edges, often "Y" shaped grow along the runners (rhizomes). One of the more prolific species and considered one of the better refugia species because of its abundant growth and being less prone to sexual reproduction disintegration. Another would be *C. taxifolia,* and possibly *C. floridana.*

Calcareous Algae

Upright growing forms of calcareous algae contain a large amount of calcium that lend a certain amount of stiffness. Growth rate depends upon the availability of calcium, magnesium and trace amounts of iron and iodine as iodide. Under the right circumstances, some forms of calcareous algae can grow as much as a few inches per day. There are numerous species, but the following three are more common and generally available:

Halimeda opuntia – Shaped like a prickly pear cactus and sometimes called Cactus Algae or Money Plant. Prefers medium to indirect light. Besides needing a calcium level near that of Natural Seawater (NSW), about 400 mg/l, it appears to do better if alkalinity is maintained between 3.5 – 4.5 meq/l or even slightly higher. When this alkalinity range is maintained I have seen this dark green plant grow in my aquariums at a rate of one-quarter inch (.6 cm) per day. It's a good species for aquariums with grazers such as Tangs because the calcareous makeup of the plant helps keep their digestive system free of blockages. *H. discoidea,* and *H. incrassata* are another two favorites.

Penicillus capitatus and *P. dumetosus* – Referred to as Neptune's Shaving Brush because it has a thick stalk crowned with a tuft of green to white filaments. Prefers sandy bottom areas and intense light. Has the same water quality requirements as Halimeda. Its upright stalks are short lived, maybe lasting only a few weeks. Yet new growths will occur within a short distance of the main stalk via runners from the disintegrating central stalk. May grow up to a few inches per day.

Udotea spinulosa – This single, broad, fan-shaped bright green leaf on a stiff, calcified single stalk is the most difficult of the three to maintain. There are various species, with height ranging from 6 - 8 inches. Prefers a sandy bottom and partially shady areas. Seems to have the same water quality requirements as *Halimeda opuntia.*

Coralline is generally an encrusting calcareous algae appearing in a variety of colors, e.g., pink, green, purple, white, yellow, maroon and brown. Most require turbulent water movement, low to medium light, broad spectrum, calcium level near 400 mg/l, magnesium near 1300 mg/l, alkalinity above 3 meq/l and small additions of iodine as iodide. The following are only a few that can show up in the aquarium:

Hydrolithon boergesenii – Encrusting; purple to lavender; spreading in knobby layers. Prefers turbulent water movement and intense light. A favorite surface for the initial growth of hair algae bubbles.

Jania rubens – Erect; rose red; straight and branching segments tightly joined by flexible joints. Tightly packed 2 inch (5 cm) clumps. Prefers low water movement and medium light.

Peyssonnelia sp. – Encrusting; dark red to maroon; edges look like old peeling paint. Prefers shaded areas.

Porolithon pachydermum – Encrusting; pinkish gray; moon crater-like, with craters sometimes caused by boring algae (chiton). An important reef builder. Prefers turbulent water movement and intense light.

Mesophyllum mesomorphum – Encrusting; overlapping/shelf-like; dark red to pink; very brittle. Found in the Caribbean and Indo-Pacific.

Sporolithon episporum – Encrusting; reddish brown; spreading in layers that overlap each other. When broken or chipped, exposed surface is white. Prefers turbulent water movement and strong light. One of the more common coralline algae. Also a favorite surface for initial growth of hair algae bubbles.

Titanoderma sp. – Encrusting; grayish-pink; found growing in small patches or groups of small 2 inch patches. Will grow on other algae, including bubble algae and Caulerpa. Prefers turbulent water movement and intense light.

Titanoderma prototypum – Encrusting; circular patterns of pinkish cream to red colored small hemispherical bumps that spread and join each other. Requires medium lighting.

Titanoderma bermudense – Encrusting; grayish to pale red; often in overlapping layers. Often with grayish lines on the surface that are the result of the sloughing away of surface cells to keep other organisms from attaching. Requires intense lighting.

Red Algae

The Division Rhodophyta is the largest group of tropical reef plants consisting of over 4,000 species. Some of these species contribute to reef building, some serve as a food source, a few can add beauty to the aquarium, and some are troublesome. Whether they can be considered beneficial or unwanted is a matter of opinion. Occasionally they show up attached to live rock such as the encrusting *Peyssonnelia sp.,* which I currently have in one of my reef aquariums. From time to time I've seen filamentous *Centroceras clavulatum,* moss-like *Wrangelia argus* and the very pretty and fairly common *Halymenia spp.* attached to newly received live rock. Rarely do they do well for any appreciable length of time.

There are also some species, such as *Dasyopsis spinuligera,* that appear to kill certain stony corals. There is still much to be learned about their metabolites and secondary metabolites and their effects on various species of stony corals. Caution is advised with unknown or little known members of this family until there is more information on these troublesome aspects.

Zooxanthellae

Over the tens of thousands of years that invertebrates have lived in the ocean depths, some have developed a symbiotic relationship with algae called "zooxanthellae." These brownish-colored (the best color for absorbing blue light) single cell algae utilize the host's nitrogen-laden waste products and carbon dioxide from the surrounding seawater to fuel their energy needs. Their waste products, sugars and amino acids, leak into the host's body where they are used as a food supply.

Ahermatypic corals lack symbiotic algae and are not considered reef-building corals. Hermatypic corals possess symbiotic algae and are generally classified as reef-building corals. Because of this symbiotic relationship it is important to provide the correct spectrum and intensity of light so efficient photosynthesis will occur. Since blue light is mainly required by zooxanthellae to trigger photosynthesis, lighting equipment should be designed so this portion of the spectrum is given high priority.

Because blue light is so important to reef keepers, I should touch on the "Growlux" lamp. For years, aquarists thought that if the Growlux fluorescent worked for their houseplants, it should do equally well for the plants and animals in their aquarium. They were right about its use in freshwater aquariums, as the lamp provides excellent spectrum for "true" plants and also enhances the colors of most fishes. Even in seawater aquariums, the Growlux lamp can provide adequate viewing conditions. Yet, it was designed to give the houseplant what it needs, a much higher red band wavelength. That is why the lamp has a reddish tint when lit. I've used and will continue to use the Growlux on freshwater aquariums, as they do a great job. But different lighting requirements are in order for marine aquariums. Keep in mind the invertebrates that interest the marine aquarist have existed in the oceans for many thousands of years at a depth that does not receive the red portion of the visible spectrum.

Unwanted Algae/Bacteria Mats

There are what I consider seven categories of problematic algae and bacteria slime-like growths that cause aquarists the most problems. They are hair algae (long hair-like/plume-like growths); slime algae; short fuzz/turf-like algae; bubble algae; brown diatom algae; dinoflagellates; and a very specific green microalgae.

Hair Algae

There are two green filamentous hair-like growths that are considered a major pest. Both are presently being considered cyanobacteria that contain chlorophyll. The first and most difficult of the two to overcome is a true hair-like growth called *Derbesia.* Its long, soft, and hair-like strands can rapidly cover wide areas. Not only does it become independent of bulk water nutrients once established, trapped detritus/debris can add further nutrients to the bulk water.

The other hair-like alga is *Bryopsis plumosa.* This very fine, feather-like and upright growing plant occurs in more bush-like growths, i.e., as a single stem-like plant with multiple, long plume-like tufts. It also becomes mostly independent of nutrients from the surrounding bulk water once the plant structure becomes recognizable.

Both *Derbesia* and *Bryopsis* plumosa have two life cycles, i.e., as a small bubble-like growth and that of an easily recognizable plant-like structure. *Derbesia* has tiny spore packages called "sporangia" that form along its hair-like strands. When these packages mature, they burst open and release spores that settle and grow into "gametophytes" which look like small green bubbles. When these bubbles mature, they release male or female germ cells that eventually unite and form a "zygote," which is the base unit for the growth of a new structure of hair algae. Though some aquarists say algae does not grow on surfaces covered with coralline, that is not true. Quite frequently the early stage bubbles of *Derbesia* can be found attached to surfaces covered with coralline, e.g., *Sporolithon* and *Hydrolithon.*

Slime-like Growths

Dark red, slime/grease-like growths are a cyanobacteria. It is photosynthetic and can cover wide areas rapidly. Most of the time it's considered a nuisance because it can get out of control "very" quickly. Some very small patches on the substrate should not be cause for alarm as it is one of the earliest microbial colonizers. Only mat-like growths should be removed.

As with hair algae, once it begins to form mat-like growths, it is capable of becoming mostly independent of bulk water nutrients. Since these mats become a barrier to normal diffusion gradients in rock or sandbeds, they negatively affect the normal microbial processes in the substrate they are covering and nearby areas. This benefits the growth of the plant structure because the conditions in the below substrate then supply most of its nutrients.

Cyanobacteria may sometimes be confused with brown "grease-like" algae, which is sometimes stringy with bubbles of gas trapped inside. This slime-like growth is usually either a very bad case of diatom alga or a dinoflagellate alga. Both can grow rapidly and coat large surface areas quickly.

Another slime-like alga in the Kingdom Monera is *Phormidium corallyticum.* It may be the cause of stony coral and gorgonian Black-band disease. Once this alga's toxic metabolite is near coral tissue, the tissue begins dying. Dead tissue then becomes a food supply for this and other opportunistic bacteria. Further necrosis spreads and as more tissue is affected, it often leads to the loss of the specimen.

Turf Algae

Turf algae has many different grass-like forms. There are finely branched delicate forms; short and stubby tubular forms; thick grass-like mats; and various other low and tall growing forms generally of the *Enteromorpha* genus. Small amounts present no problems in most aquariums. In some ways it can serve as a food supply for herbivorous animals. However, it can become a problem if not controlled. Combine its initial rapid uptake of nutrients with their large amount of surface area and the result is quick growth. When turf algae becomes well established, it can become a collection device for detritus, adding to bulk water nutrients.

Bubble Algae

Bubble algae have become a problem over the past few years. The various species of *Valonia,* or *Ventricaria ventricosa* are oval, spherical, or tube-shaped green single cell algae. They mostly attach themselves in rock crevices or depressions where nutrients tend to collect. They grow well under various lighting intensities and even do exceptionally well in aquariums with very low nutrient levels.

Their bubbles can grow fairly large and I have seen some as large as 2 inches. When their membrane-like wall material wears out/ruptures, internally forming new cells/bubbles are released and can rapidly spread throughout the system.

Brown Diatom Algae

Probably the most common brown algae are the brown diatoms. They are usually one of the aquarium's most common/first colonizers. This type algae can be sheet/film-like in appearance, appear as tiny brown dots on the aquarium inside panels, or cover the substrate surface where there is sufficient light. Silica/silicate is used in the structure of their cell wall. Depending upon the amount of silicate in solution, their growth can appear dot-like or mat-like. A small amount of this algae adds cleaning time to the general maintenance schedule but will not harm the animals in the aquarium. Besides appearance, when heavily coating objects it can photosynthesize and produce oxygen, often seen as bubbles, which in turn might raise pH to dangerously high levels (above 8.6).

Dinoflagellates

Another type of problematic slime-like brown algae is caused by a dinoflagellate. It can result in large areas, including the surfaces of invertebrates, being coated in a brown gelatinous mess. It is usually introduced on a new piece of live rock or coral animal and can easily become problematic if the bulk water is rich in organic material.

This free-swimming algae is extremely difficult to cure. Silicate is not one of its needs as is the case with diatoms. Unfortunately calcium carbonate is. Since we cannot rid the aquarium of its calcium carbonate content, the various methods described in "Remedies" should be employed. Bear in mind that a small microscope would help identify which form of brown algae is in the aquarium. If the alga cell has a flagellum, i.e., a tail (for propelling itself through water), it is a dinoflagellate.

Green Microalgae

There is a particular parasitic form of green microalgae that can occur in the calcium carbonate skeleton of stony corals. This boring microalgae, *Ostreobium spp.,* grows inside coral skeleton material. It reduces the structural strength of the skeleton material and interferes with the animal depositing new skeleton material. Any stony coral that may exhibit a greenish coloration in its skeleton should not be purchased, as curing the problem area is doubtful.

Contributing Factors

There are a number of factors that lend themselves to the beginning or continuance of nuisance algae and slime-like bacteria growths. Some of these factors are what could be considered having a more indirect than direct effect. Yet all influence growth sooner or later. Let's look at all reasonable nutrient sources, beginning with that all-important solvent – water. Then look at the actual nutrients themselves along with other contributing factors.

Water Quality

Even though most aquarists spend much time on planning the various building blocks of an aquarium system, such as lighting and filtration equipment, they sometimes give little thought to this medium except for salinity. Monitoring tap water can be helpful in minimizing certain nutrients, e.g., nitrate, phosphate and silicate that can enter the aquarium in this manner. Sounds simple, but it can be somewhat more complex than just testing the effluent that comes from the tap. In fact, long before filling the aquarium with water, the quality of the tap water should be checked as it simply doesn't make sense to blindly fill the aquarium with untested water.

The United States Environmental Protection Agency (EPA) has defined quality requirements for water delivered by municipal water sources. Most public water supplies have contaminates well below EPA levels and some of our more delicate aquariums have done fine on public water supplies. For instance, the EPA's standard for Nitrate (as $N0_3$-N) is 10 mg/l. This is slightly below the suggested maximum level for reef aquariums, which is 15 mg/l. Heavy metals such as copper are extremely toxic to invertebrate. Yet, there's not much concern when it comes to human consumption. Furthermore, almost all municipal drinking water in the United States is treated with chlorine or chloramine (a combination of chlorine and ammonia), to kill disease carrying organisms. To be on the safe side when delicate specimens inhabit an aquarium, wherever you may live, it's recommended some testing/processing be performed on the water that exits your tap.

Tap Water

In most locations, plain tap water and a quality salt mix can be used for starting a new aquarium, or used directly for water changes or evaporation make-up. Nonetheless, it should first be tested for the nutrients mentioned above, and especially copper if it's for an aquarium containing invertebrates. Also, the water company supplying your locale should be contacted to see if they are using a phosphate-based rust inhibitor in their delivery system. Bear in mind the water company customer in the United States is entitled to a free test of the water coming to their home or business per federal law. If not positive about its quality for you or your aquarium, discuss scheduling a test with the local supplier or at a minimum, test it yourself.

As for copper, if tap water is going to be used for the fish-only aquarium, it should contain less than .2 mg/l copper. Keep in mind that some fishes are sensitive to extremely small amounts of copper, e.g., Angelfishes, Butterflyfishes, and Puffers to mention a few. So care needs to be taken if any copper is present in the incoming tap water. Of course, if used for the reef aquarium, no copper should be present.

If the tap water needs to be processed, it can be flowed through activated carbon or an ion exchange molecular absorption filter pad until the copper level is within acceptable limits. Or, the tap water can be processed through Reverse Osmosis (RO) and/or Deionization (DI) equipment as explained below. Many times, simply allowing tap water to run for a few minutes will either greatly lessen or totally remove any copper ions present. The reason for that is many homes or businesses have copper piping. Stagnate water accumulates a small amount of this element from the piping. Flushing the pipe simply brings in a new supply of water that has not yet had the time to accumulate the copper ions.

When it comes to silicate, I find it best if tap water contains less than .2 mg/l. Usually, tap water throughout North America is fairly high in silicate and can generate diatom algae as noted above. To lessen silicate, the RO process is fairly effective. Yet, the DI process is much more effective.

As for nitrate (NO_3-N), any amount above a few mg/l should be a signal to process the water through RO and/or DI equipment.

Phosphate is a major plant nutrient and care needs to be taken to introduce the least amount possible, e.g., <.05 mg/l. Many water suppliers periodically introduce a polyphosphate or a zinc orthophosphate compound into the water supply to reduce rust in old pipe systems. Aquarists usually find out about these rust inhibitors when it's too late, i.e., when an algae bloom unexpectedly shows up in their aquarium. I recommend occasionally checking with the water supplier. Again, RO and/or DI equipment become a vital link in providing quality water for many aquariums.

Well Water

Some folks still utilize well water in their homes. The number of wells for the purpose of decontaminating polluted groundwater supplies is growing in North America, whereas the number of wells for drinking water is diminishing. Most drinking water wells are found near farm country where pesticides, nitrate and phosphate fertilizers have had the opportunity for many years to percolate downward and reach the aquifer. In some areas of the United States, well water is no longer safe and farmers must depend upon surface water to irrigate their crops or for drinking.

The wells that are still being used for drinking water should be checked yearly for contaminates. Even if they meet EPA drinking water regulations, the water should still be checked for plant nutrients, i.e., nitrate, phosphate, silicate, and possibly iron before used in the aquarium. Bear in mind that even though well water may meet all standards, it is usually low in oxygen and almost always high in carbon dioxide. Therefore it should be aerated for 24 hours prior to use in the aquarium.

Reverse Osmosis

This water purification process removes about 90% of dissolved salts and organic compounds. The process was first discovered being used by seabirds. It was noticed seabirds would drink seawater and appeared to be unaffected by its high salt content. Upon a closer look it was found the birds actually filtered the seawater through a membrane in their beak, resulting in only freshwater being taken into their system.

There are two types of membranes used in RO equipment: cellulose triacetate (CTA) and thin film composite (TFC). The TFC membrane is far superior in that its rejection rate of pollutants is much higher, about 90 to 95% depending upon the particular pollutant. It also does not readily break down from bacteria attack.

Since the membrane is the heart of the RO system, it must be protected from certain impurities or disinfecting solutions in the incoming water supply. To protect the membrane from coarse solids that would quickly clog its very small holes, water should flow first through a Sediment Prefilter. Next, an in-line filter should be installed with activated carbon or preferably a carbon block as a filter medium. TFC membranes are very sensitive to chlorine and since water supply companies generally use chlorine to disinfect water lines, at least in North America, the carbon prefilter is a must. Besides removing chlorine and pesticides, some other volatiles will also be removed.

Water then flows to the membrane where, via the water pressure in municipal water system pipes, only the very small almost pure water molecules can pass through the membrane holes. Larger contaminant molecules cannot fit through the tiny membrane holes and are rejected via a drain line. Of the three main plant nutrients, the nitrate and phosphate anions are easily rejected. Yet, all RO units have difficulty rejecting silicate for any appreciable length of time.

The only drawback most of these units have is they only provide one usable gallon of water for about every three to five that flows through them. Yet, it seems like the more water rejected, the better the membrane rejects pollutants. However the rejected water is still usable for watering house/garden plants or for use in Cichlid aquariums.

No matter what brand RO is chosen, there should be individual replaceable sediment and carbon prefilters along with a quality brand TFC membrane. Pay special attention to the manufacturer's replacement schedule for the sediment, carbon and membrane filter. Depending upon usage, most carbon filters need to be replaced every six months, and the sediment filter and membrane every one to two years.

The general aquarium RO units are rated in how many gallons they will produce in a twenty-four hour period. Usually the 10 to 50 gallons per day units will suit most aquarium needs. Water pressure is another important operating parameter and 45 pounds should be considered minimum; 55 pounds very good; and 65 pounds ideal. A quality made RO system should be considered a worthwhile investment for most fish-only and reef aquariums.

Deionization

This type of equipment uses resins that attract unwanted compounds and elements, thereby selectively removing them from the water that passes through the unit. DI units, or IX equipment (Ion eXchange) as they are sometimes called in the trade, are sold in two different styles – having a mixed bed of both anion and cation resins in a single container or with separate beds of each type resin, each in its own container. The anion resins filter out negatively charged ions and the cation resins remove positively charged ions.

There is no wasted water as there is with RO units. What passes through the DI unit is 100% nutrient free and totally usable. Now don't go out and buy a DI unit and then throw out your RO unit. It is better to first process water through an RO, then pass it through a DI unit. Why is that you may ask? Water from the tap is usually high in many of the anions/cations that mixed or separate bed DI units remove. When these beds become clogged with all these unwanted ions they either need to be recharged or replaced if a mixed bed unit. Not pleasant for your wallet or for performing needed maintenance! By first removing the majority of these unwanted ions with an RO unit the time frame between DI recharges/ replacing cartridges will be greatly lengthened. Keep in mind only a strong lye solution can be used to recharge an anion resin and only acid can be used on the cation resin. Therefore only separate beds can be recharged. Much care needs to be taken when using these dangerous recharging solutions and their disposal when the task is complete. Mixed beds are much more user friendly!

Distilled Water

Many aquarists use distilled water thinking that if it is okay for the battery in their automobile it must be pure enough for their aquarium. Even though distilled water is formed from condensation, the condensate, i.e., the water collected, could possibly contain trace amounts of copper if some of the equipment used in the distilling process is made from copper. It is always wise to test distilled water for this toxic, heavy metal prior to its use.

Nutrients

The processing of freshwater is a good procedure to consider if closed systems are to remain more algae free. Another logical step to consider would be the quality of the seawater already in the aquarium. It may have been a highly suitable solution when first placed in the aquarium. Nevertheless, as the system ages and bio-load increases, water quality can easily diminish. Knowing what constituents are considered prime plant nutrients, why to test and how to correct if necessary, is important in maintaining on-going water quality if a mostly algae free environment is the goal.

Phosphorus/Phosphate

Some aquarists may be confused with the utilization of the word phosphorus (P) and/or phosphate (PO_4). Actually, phosphorus is the source material for the production of phosphate, a combination of phosphorus and oxygen, by microbial processes. Phosphorus is an important, in fact a necessary building block of all life as it is part of "every" DNA molecule. It enters the ocean from the land and undersea volcanoes as Dissolved Inorganic Phosphorus (DIP). There are two other forms of phosphorus: Dissolved Organic Phosphorus (DOP) from ruptured algae cells, animals, and bacteria, and Particulate Organic Phosphorus (POP) as part of detritus and other organic matter.

Phosphate/orthophosphate (DIP) is the most important of the three. The anion phosphate, which is a major nutrient for algae growth, can also interfere with the growth of some corals. Yet, it is needed in minute quantities by animals, including symbiotic and coralline algae. In fact, it may be why early stages of hair algae colonize certain types of coralline – to gain access to phosphate so as to provide energy for initializing growth.

If aquariums had the same processes as the ocean, phosphate would almost always be below detectable limits using the average aquarium test kit. These kits generally have a low reading of .05 mg/l and the ocean has a concentration of approximately .03 mg/l.

There are several ways in which phosphate enters the closed system. Tap water is one of those ways because phosphate is used as a fertilizer on many farms. It tends to percolate downward and enter the aquifer, or washes off the surface and enters nearby rivers and streams. It then finds it way into local water impoundments and eventually flows from private or municipal water suppliers to the tap. As noted above, some local water companies also use phosphate as an additive to control rust in metal pipes/fittings. Getting an analysis of the incoming water to your place of business or home, or sampling it yourself is a good first step to take before using it in the aquarium.

Another way phosphate enters the aquarium is from the mineralization of dissolved and particulate organic animal/plant waste products by heterotrophic decay bacteria. Phosphate may also enter the aquarium when activated carbon is used, including those that are acid washed. Some simply leach more than others. For details on how to select the right activated carbon see my *"Protein Skimming and Activated Carbon Secrets"* booklet. It's also possible for certain substrates used in calcium reactors to introduce phosphate. Low pH in reactors makes it easy to dissolve this nutrient, often bound to the reactors' calcareous material, especially if it's shell-type material. Bear in mind that shell-type animals consume algae/filter feed on organisms containing algae/heavy metals. It is then incorporated into their shell only to be leached back into solution by the calcium reactor's low pH.

Not only is dissolved phosphate a nutrient for plants in general, it can also combine with the calcium ion and precipitate out of solution as calcium phosphate. Unfortunately this calcium phosphate adheres to anything in the aquarium that is calcium carbonate. This nutrient "coating" provides a fertile area for unwanted algae to get a foothold. This coating of phosphate is similar to throwing a handful of fertilizer on the lawn and watching the grass grow well in just the area where the fertilizer was applied.

Phosphate also interferes with the growth of stony corals and can result in tissue receding from their calcium carbonate skeleton. Since there is no way or any reason to completely eliminate phosphate, the question becomes just how much is too much. In the reef aquarium environment there is no doubt it should be kept below .05 mg/l. In fish-only systems, anything above 1.0 mg/l should be a signal to review husbandry practices. There are two phosphate removing mediums readily available; an aluminum oxide material, and iron hydroxide impregnated pads.

As for aluminum oxide media, it's capable of lowering phosphate to .02 mg/l levels. Phosphate becomes irreversibly bound to aluminum oxide and will not be released back into the water. Aluminum oxide will also weakly attract carbonate, initially somewhat reducing alkalinity. Yet, since phosphate will replace the carbonate on the aluminum oxide particle, the carbonate will return to solution. Aluminum oxide will also remove silicate, which is a diatom nutrient. Organic acids are also removed and in highly polluted aquariums this media should be replaced almost weekly. In well-managed aquariums, media replacement need only occur every 6 - 8 weeks. One further note, the product can cause some leather corals *(Sarcophyton spp.)* to close up and remain closed for long periods. In some cases, they will never come out of this closed state and will slowly waste away.

The iron hydroxide impregnated pads are advertised as powerful enough to actually cause calcium phosphate precipitant to re-enter solution where it can be attracted to the iron on the pad. Unfortunately these pads are fairly expensive and short-lived. Yet, pound for pound so to speak, they are many times more efficient than aluminum oxide media. Occasional use of these pads should keep phosphate levels in check in most systems. In situations where excessive phosphate is troublesome, use of these pads is highly recommended. There are three brands that I know of, all equally excellent products.

Generally speaking, I recommend the use of a phosphate removing media as soon as the aquarium is up and running. If the aquarist waits until he or she sees a positive test kit reading, some phosphate has already precipitated and may just be creating a fertile area for future algae growth. There are numerous ways to limit the amount of phosphate and they are noted further below. However, water changes will not be one of them. The reason for this is the majority of phosphate in the aquarium is probably bound to anything that is calcium carbonate. Therefore a water change, even a large one, is so temporary that it may last only a day before the precipitated phosphate begins leaching back into solution.

Keep in mind that aquarium test kits check only for inorganic phosphate. Organic phosphate, i.e., phosphate tied to detritus and other organic substances, will not register on an inorganic phosphate test kit. Therefore it's possible to have greater amounts of phosphate than what registers on these type of test kits. How often should phosphate be tested depends on how well the system is maintained. I would recommend testing weekly until the system is well established. Then monthly should suffice unless some major bio-load changes are made.

Ammonia

Once tap water is processed, if necessary, and mixed with a salt mix it should be tested for ammonia (NH_3) before being placed in the aquarium. You'd be surprised as to what may be found! As for the newly established aquarium, ammonia is the first chemical compound tested during its nitrification cycle. Unless there are major increases to bio-load in its future, ammonia will probably be tested infrequently thereafter.

Most aquarists do not realize ammonia is probably the number one fish killer! (And you thought it was people.) Ammonia is so dangerous that it can burn gill membranes to the point where permanent gill damage results. Internal organs can be damaged; blood vessels can hemorrhage; eyes can be burned or scarred; stress levels increased; nerves damaged; and, the production of fish slime, an important protection against parasites, shut down. That's besides being a primary algae nutrient! These are all good reasons not only to monitor ammonia, but understand more about this hazardous compound.

When protein molecules break down, nitrogen (N) and hydrogen (H) combine to form compounds called ammonia and ammonium (NH_4). When ammonia is discussed, we are generally referring to "total ammonia" or the sum of free ammonia and ammonium, both in solution at the same time. The less dangerous ammonium ion is more commonplace at or near a neutral pH. Yet, as pH rises, ammonium is quickly converted to the more toxic free ammonia.

The concentration of free ammonia increases about 10 fold when pH moves from 7.0 to 8.0. Between 8.0 to 8.3, it can, depending somewhat on temperature and salinity, increase another 10 fold. This simply means a marine aquarium with a given amount of ammonia, would be safer at a lower pH, lower temperature, higher salinity, than the same aquarium at a higher pH, higher temperature and lower salinity. Whether in new or well-established aquariums, it is wise not to deliberately raise pH too quickly as it affects ammonia equilibrium much more than does temperature and salinity levels. It's another consideration the users of Kalkwasser should carefully consider if they are not using carbon dioxide injection equipment to lessen its impact.

Generally, most ammonia comes from animal waste products when released into the aquarium water. Most of the remaining ammonia content comes from microbial decomposition of organic matter. Having an adequate biological filter, correctly sized to existing bio-load, will go a long way in keeping "free ammonia" at very low levels. However the not-thinking homemaker can create a source of "instant" ammonia in the aquarium. This would be the person who goes around the home using an ammonia-based spray cleaner, happily cleaning all windows, counter surfaces, etc., in sight. Ammonia-based household cleaners contain "free ammonia." Any aquarium in close proximity can directly absorb the vapor from these sprays. Forewarned is forearmed!

Some aquarists think that testing for ammonia once the nitrification cycle is complete is a waste of time. Not true. Besides presenting a dangerous condition for animal life, ammonia is the exact food algae require for growth (phosphate helps provide the energy needed to utilize it). Therefore, occasionally testing for ammonia is not as far-fetched as some aquarists tend to believe. Because of that I'm including some testing reminders:

- During start-up of the new system and prior to adding fishes or invertebrates.
- Prior to and after adding to the current bio-load and that includes live rock whether cured or not.

- Testing the freshwater used for a water change or evaporation make-up.
- Testing newly prepared seawater as some salt mixes generate ammonia.
- Testing the aquarium prior to a large water change, especially if a change has not been accomplished for a long time.
- Right after accomplishing some household chores that required the use of strong cleaning agents.
- When first noticing one of the thought to be current inhabitants not visible.
- After adding a medication.
- Before adding to or modifying system equipment.
- And, simply every few months in well-established, stable environments.

As for ammonia test kits, there have been many improvements over the past decade. I've even tried the dry strips and for a quick check, found them reliable. Keep in mind we are in need of a test kit where the range of interest is 0 to no higher than 1.0 mg/l and where .1 mg/l increments can clearly be distinguished.

As to just what level of ammonia is lethal, well, that's like asking how much exercise would harm a person when you know nothing about their health. There are simply too many variables to make an accurate table depicting toxicity levels. However, it's safe to say that any readable level of ammonia should be cause for concern. Levels of .1 mg/l can be lethal for some animals. Others can withstand higher levels. But none can continue for long without some kind of damage or continue to process and dispose of their ammonia/ammonium laden waste products against the back pressure caused by "ammonia" in solution. Just keep in mind that any sign of ammonia in a well-established aquarium is like smoke rising above the forest. Where there is smoke, there's fire! Where there is fire, there's danger. And, algae find it less energy-intensive work to assimilate it directly than generate it in their internal processes.

Nitrate

About twenty years ago, very few aquarists were concerned about nitrate (NO_3) levels. Once the emphasis on reef aquariums began, somewhere about the mid-eighties, nitrate became a compound of major interest. Reef keepers quickly became aware that in their specialized ecosystems, excessive levels of this compound could cause stress or the demise of their prize invertebrate specimens and help generate unwanted forms of algae. The aquarist began asking what is nitrate, where does it come from, why should it be of concern, when to test for it and how to minimize it.

As for the nitrate ion/molecule, it is a combination of one nitrogen (N) atom and three oxygen (O) atoms. It is the end result of the nitrification cycle, i.e., ammonia oxidized to nitrite, then to nitrate. For a very long time many aquarists considered nitrate to be harmless. Levels well over 100 mg/l appeared not to bother most fishes. Some fishes, such as lionfish, damsels, triggers and moray eels easily withstood even higher levels without exhibiting the slightest sign of discomfort.

As more scientific evidence became available it showed that in certain circumstances, nitrate could be detrimental to both fishes and invertebrate and that it was an algae nutrient. Some studies reported that concentrations as little as 20 mg/l could cause high mortality rates in certain types of fish fry or their eggs. Various other articles pointed towards some invertebrate having their respiration affected by levels of nitrate between 30 to 100 mg/l. Yet, on the whole, there is still not a lot of well-defined scientific data on how nitrate affects the animal species most aquarists keep in their aquariums. But my Caulerpa experiment as noted above, clearly depicted its value to the alga.

Regardless, most current information points toward keeping nitrate in the reef aquarium at a much lower level than we would allow in a fish-only aquarium. Information in my personal library points towards not exceeding 15 mg/l in the reef aquarium. For further reading on this subject, review *"Seawater Aquariums, The Captive Environment"* by Dr. Steven Spotte. If serious about this hobby, it's a book that should be in your library. In fact, after years of experimenting with reef aquariums, I am convinced best results are achieved by staying under 5 mg/l NO_3-N.

I recommend testing a newly established aquarium before stocking. Should the nitrate levels be higher than 5 mg/l, a major water change should be considered. While adding to the bio-load, I suggest testing once a week, then monthly after the system is well established so as to have an idea if improving general maintenance procedures is required.

As for reducing or controlling nitrate accumulation in the established reef aquarium, water changes are basically an incorrect approach. This may be fine for the newly set-up system where nitrate may be somewhat high from the initial cycling period. But for long-term nitrate control it is not a cost-effective approach. It would simply take massive water changes on a steady basis to keep a serious nitrate problem within an acceptable range. Bear in mind that in reef aquariums, nitrate accumulates inside live rock or in the depths of a sandbed. Using water changes to reduce nitrate levels may seem effective the first day. However, nitrate will flow out of the rock and sandbed interiors, back into the aquarium's bulk water over the next day or two. Within a day or two the aquarium's nitrate level will be almost back to the level it was just prior to the water change. It makes more sense to get to the root of the problem.

There are two different methods for reading nitrate. Some nitrate test kits measure the "nitrate ion" and others measure "nitrate nitrogen." Nitrate is a compound, or combination of elements. One molecule of nitrate is composed of one nitrogen atom and three oxygen atoms. Since the atomic weight of nitrogen is 14.01 and the atomic weight of one oxygen atom is 16, the weight of one nitrate molecule equals 62.01, or 14.01 + 16 + 16 + 16. Therefore, the test kit that reads the nitrate molecule will show a reading 4.4 times higher than a test kit that reads nitrate-nitrogen. I should note reader letters frequently fail to mention whether their nitrate level was the nitrate molecule or nitrate-nitrogen NO_3-N reading. Even mentioning the brand test kit would be helpful. Note, when I refer to nitrate levels in my writings, it always relates to nitrate-nitrogen unless otherwise noted.

As for NSW, it has very little nitrate, e. g., $N0_3$-N = .03 mg/l at or near fringing reefs where much of our prize hard corals come from. In and around shoreline areas, nitrate readings rise slightly and certain kinds of invertebrates, such as soft corals, exist without problems. As noted above, it's unfortunate there are no specific scientific studies on just what levels of nitrate (either nitrogen or the molecule) affect the average invertebrate we keep in our aquariums. But experience is a good teacher and as noted above, I recommend nothing greater than 5 mg/l nitrate-N for the average reef/invertebrate aquarium. Yet, for the reef aquarium containing more delicate small polyp stony (sps) corals, my "goal" is less than 2 mg/l.

I should also note that nitrate is not affected by ozone. Ozone can oxidize ammonia into nitrite, then into nitrate, especially at a high pH, but no further. Moreover, at the pH we maintain our aquariums, 8.1 - 8.3, ozone probably has much less "conversion" power than do nitrification bacteria.

Keep in mind the level of nitrate is an indicator of overall microbial balance. If the quantity of microbes that normally and effectively use nitrate is equal to or greater than the microbes that produce it, the system has achieved balance through equilibrium. If nitrate accumulates, the aquarist must recognize where the system defects are and be willing to correct them. Some are easier than others to correct. However without a good microbial foundation, system corrections may be expensive and time consuming. If not adequately resolved, unwanted forms of algae will benefit.

Silica

Silicon (Si) enters the oceans in a similar fashion as does phosphorus. When it combines with oxygen, it becomes silica (silicon dioxide (SiO_2)). In a dissolved form, e.g., Silicic Acid (H_2SiO_3), it quickly simulates diatom growths, which use silica in the structure of their cell wall. Even though silica, as silicate (SiO_3^{2-}) varies greatly in NSW, e.g., .04 - 8+, anything over .2 mg/l in closed systems may cause a diatom bloom. Silicate is usually introduced into the aquarium via tap water, a salt mix, silica sand dissolving slowly (possibly in fluidized bed filters), or something in the aquarium that may be composed of a silica substance.

Tap water even if processed through an RO system will still contain silicate. Very few RO units are effective at removing silicate. Even quality RO units remove only about 90% of the incoming silicate. To remove all silicate from tap water, one must utilize the proper combinations of resins in a well-engineered DI unit. Salt mixes should also be tested prior to usage no matter what is read in their advertisements as silica may be used as a drying agent. Anything that may be composed of silica, has a slight solvency at the pH we maintain our marine aquariums. In fact, silica starts to go into solution at a pH of 8.0. It greatly accelerates its solvency at a pH of 8.5. It is wise to at least test incoming tap water and newly prepared seawater for this nutrient. Once in the aquarium, it is used quite rapidly and a very low aquarium level can be quite misleading.

Iron

Iron (Fe) is a minor trace element found in natural seawater and is necessary for algae biological oxidation/reduction and chlorophyll production along with animal/coral coloration and growth. How much is required for these functions is difficult to quantify. I have seen recommendations that .1 mg/l level of iron be maintained in aquariums at all times. Yet I think that was by someone who was selling an iron additive. A constant level of .1 mg/l doesn't seem to make sense since the level of iron in NSW is only .01 mg/l.

Whether or not to add or test for iron, unless the desire is to have lush growths of macroalgae, is much debated. In my opinion, it's not necessary to test or specifically add iron supplements. However, if you're using well water or live in an area that naturally has high iron deposits, you may want to at least check the incoming water for this nutrient. Keep in mind that unless iron additives are chelated, they will quickly precipitate out of solution at a pH of 8.0 and above.

Another factor, especially notable for reef aquariums, is that iron fosters release of phosphate bound to calcareous material in areas subjected to the sulfur cycle. Since the sulfur cycle is a quite normal microbial process in all sandbeds, there's absolutely no good reason to specifically supplement iron in reef aquariums that contain any type of sandbed. In fact, if sensible water changes are accomplished, the salt mix should fulfill all necessary iron needs.

Molybdenum

Molybdenum (MoO_4), found in NSW at .01 mg/l, is a trace element that is essential to nitrogen-fixing organisms, e.g., cyanobacteria (red slime algae) and possibly zooxanthellae to some extent. However, excessive use of molybdenum, pronounced ma-lib-de-num, in conjunction with phosphate (including DOP and POP) and sufficient light is a major algae/slime algae enhancer. This is another trace element additive that I don't believe is needed. Any requirements for it can adequately be met via water changes with a quality salt mix.

Water Movement

One of the most overlooked parameters in both fish-only and reef aquariums is thorough movement of the medium itself. Most aquarists realize water movement is a way of bringing oxygen to the animals in the aquarium and reducing the influence of carbon dioxide on the carbonate buffering system. However there is often insufficient water movement and that can lead to detritus accumulation. Detritus then breaks down adding nutrients to the surrounding water or settles in crevices where algae of one kind or another can get a foothold. Increasing water flow with surge devices, additional or large powerheads not only helps to eliminate low flow areas that favor cyanobacteria, but also helps suspend detritus long enough for mechanical filtration to remove a portion of it.

Light

Some aquarists blame the quality of their lights for algae growth. Others blame copious amounts of dissolved nutrients. There's even the thought that new lamps, which have greater intensity than older lamps, may enhance algae growth. However, none of these are totally true. Light alone cannot cause the growth of algae. Nutrients alone cannot cause the growth of algae. But combine them and there will be algae growth. Furthermore, if the aquariums bulk water contains copious amounts of skewed constituents, whether that is so-called nutrients or excessive amounts of additive products, even additional quality spectrum and intensity will enhance algae growth! The reason for this is that skewed bulk water element levels are taking the blue-green spectrum waveband coming from the lamp and changing it into a more red band wave length that algae prefer. The point here is that light alone can not be blamed for unwanted algae growth. However, quality light spectrum, i.e., in the blue-green band wave, can shunt the growth of algae "if" the water it is passing through is more like that of natural seawater.

Live Rock

One of the most overlooked causes of initial algae problems in some new aquariums is the use of poorly collected and transported live rock. Many organisms that live on or inside live rock die when subjected to air. The resultant nutrients are simply more than a newly established system can adequately process.

Selecting live rock at a local dealer gives the buyer not only the opportunity to get just the type and shape rock he or she wants, but also rock that has had an opportunity to dump some of its nutrients into the dealer's tank. (Better the dealer's tank than yours!) But the same may not be true when purchasing directly from the collector or through mail order. If procuring through the mail, what the aquarist does with it upon taking it out of its shipping container will be the most important care that rock "ever" receives! Even if receiving quality seeded/cured rock, I recommend some cycling, if possible, before it enters the aquarium. Planning ahead and being ready for the shipment before it arrives simply makes sense.

If the aquarist needs to take rock from the shipping container and place it directly in the aquarium, purchasing branching or cured/seeded live rock is the best way to go. Keep in mind that even if the rock comes out of the shipping container smelling ocean fresh, there has been some die-off due to lack of water and dissolved oxygen during shipment. It is better putting the organic load generated by that die-off, no matter how small it may be, in a separate cycling container than in the aquarium. To denote the differences between cured and seeded rock, I'm including the following descriptions:

Cured Rock – This is rock the collector has made an additional effort from the time of collection to the time of shipment to keep wet. They may also scrub the rock clean of unwanted algae and/or sponges. Some collectors that ship direct to the aquarist may keep cleaned rock in a submerged holding area for several days allowing for unwanted nutrients to leach out. Rock is then shipped covered with wet newspaper, but not submerged in water so as to save shipping costs. Some of the better dealers inspect and again clean the rock as necessary before shipping to their customers.

Seeded Rock – This term has been long used to describe "cured" live rock that may have had additional time after initial collection and cleaning so as to have bacteria fully re-colonize (seed) its surface and interior areas. Some dealers like the term, some don't. Yet, I look at it as rock that has had the time and effort applied to it so as to render it mostly nutrient free and ready for the aquarium.

As for cycling mail order live rock, if your budget limits selection to other than already quality-cured rock, you definitely want to first cycle it before it goes into the aquarium. To do that, set up a plastic garbage can half filled with freshly prepared seawater a day before the shipment arrives. How many and what size containers depend upon how much rock was ordered. Place a large powerhead for circulation in the container and if possible, hook up a protein skimmer. Within five to ten days most rock should be devoid of a large percentage of its excess nutrients and be ready to place in the aquarium. This may seem somewhat labor intensive and maybe not worthwhile. But experience has taught me that it's better to be cautious and thoughtful in the beginning than sorry in the long run!

Aquascaping

Whether it's a fish-only or reef aquarium, it behooves system maintenance and long-term water quality if the rock is not placed like they were bricks! Under no circumstances pile the rocks up against the rear wall of the aquarium. Maintenance of that area will simply be impossible. Use your imagination and try to duplicate nature, not a brick wall or a pyramid-shaped structure where detritus would collect. Also, not only can too much live rock restrict water flow, it provides a lot of anaerobic area where inefficient denitrification occurs. A combination of too much and poorly placed rock can quickly lead to diminishing water quality and sustained growths of unwanted algae.

Filtration

Many new aquarists consider filtration as a process that renders crystal clear water. Actually, it should be considered a process that renders "quality" water. The only way to attain quality water is by becoming well educated with different forms of filtration "before" the system is established.

Biological, chemical, and mechanical filtration each add their own unique contribution to the overall health of a closed system. If properly established and maintained, they act together as a team and provide a balanced environment. It's these filtration methods that are the "foundation" of every closed system!

Generally speaking there are very few aquarium environments that can function long-term solely on biological filtration, even if it's extremely efficient. Chemical filtration, e.g., protein skimming and activate carbon processes, effectively remove plant nutrients or soon to be nutrient compounds before they can be acted upon by microbial processes. Of course, these chemical filtration methods are not a substitute for proper biological filtration and continued water changes. However, I consider them something like the shock absorbers in a car – nice to have for those bumps in the road. Would you like to drive a car without shock absorbers? It doesn't make sense not to use these good tools early on in closed systems. Waiting until there's a problem, then adding these tools to the system is illogical!

As for mechanical filtration, removing large matter that would deteriorate and add further organic compounds to the bulk water makes sense. Removing minute matter is usually unnecessary unless a specific problem can be resolved. And, in reef aquariums where much of this minute matter can serve as a foodstuff for filter feeders, overfiltering is a poor husbandry practice.

Overcrowding/Overfeeding

Even though lack of microbial equilibrium and balanced filtration processes are the primary causes of unwanted algae, overfeeding and/or overcrowding should be considered contributing factors. Algae require nutrients, e.g., nitrogen-laden compounds, and light to get started. In crowded or overfed closed systems the nutrients required for algae growth can easily be attained. In fact, not only easily attained, but they also tune the spectrum to something better suiting the needs of algae!

When animals are fed, usually fishes, there is generally a need to feed a balanced diet of carbohydrates, fats and proteins. Carbohydrates and fat are converted to energy and neither contains the element nitrogen. Therefore they do not contribute to the accumulation of nitrogen products in the aquarium. However proteins are a horse or fish of another color, as they contain nitrogen. During the process called "metabolism," the nitrogen element is incorporated into a compound called ammonia. When animal waste products are introduced into the water, they are basically ammonia-laden compounds. Since aquarists can't do away with foods containing proteins because they are broken down into amino acids and used in creation of new tissue for growth, thank goodness for certain species of bacteria.

Ammonia-laden compounds are used by some microbes as a foodstuff and are thereby processed into less toxic compounds. However, if the incoming supply of microbial foodstuffs exceeds the capability of the existing mass of microbes, storage occurs, usually seen as algae. The more animals to feed, the more foodstuffs in one form or another enters the system.

There are some aquarists that think their fishes need three square meals per day. I would agree some fishes such as angelfish and tangs fare better with numerous small feedings per day. Nevertheless, in the wild, fishes need to work very hard at getting their required nutrients. They may go long periods without finding proper nourishment. When food is available, their metabolism is excellent and their waste products are usually void of unprocessed food. Yet in the aquarium, fishes are certainly not getting the same level of exercise as in the wild. Over abundance of food from the over generous aquarist can certainly lead to poorly metabolized waste products. This simply contributes to poor water quality. It's better to understand the nutritional needs of the animals, then meet those needs, but not exceed them. And, of course, uneaten food only hastens diminishing water quality and often initiates slime-like growths.

Herbivores

Natural reduction of unwanted algae, sometimes referred to as "biological control," has great value if handled correctly in the closed system. If not, it adds to the nutrient load furthering unwanted growths of algae.

In the wild, herbivores concentrate their efforts in areas where algae growth is plentiful. In some areas of the Sea of Cortez, I've seen urchins by the thousands come out in early evening and begin scouring the rocks in the shallows. Much of their waste products are quickly broken down by turbulent water movement and dispersed into the bulk water. Even though their waste probably has zero impact on the surrounding water, continued algae growth in the area reoccurs because of sewage discharge from nearby hotels.

In the aquarium, waste products from large amounts of herbivores, unless siphoned out, adds further nutrients to the system. A system already skewed in favor of algae growth would therefore only see temporary results. There's no doubt reasonable use of some herbivores is helpful in closed systems. Yet in systems with prolific algae growth, it makes more sense in my opinion to limit the avenues for algae growth than spend money and time trying to control it.

Of course that may not always be feasible, and if herbivores are required, it is wiser to remove any readily accessible growths by hand before allowing them access. The less they have to turn into fecal pellets, the better off the entire system will be. Keep in mind that once fishes, including herbivorous fishes, become accustomed to a free meal, i.e., being hand fed different foods, they become less apt to vigorously devour algae growths in the aquarium.

Since there are herbivores that eat only certain forms of algae, I've created six, broad-ranging categories. A number will appear in front of the herbivores listed below indicating what general category of algae they consume.

1 – Filamentous

2 – Macroalgae/Micro/turf algae

3 – Coralline

4 – Cyanobacteria

5 – Diatoms

6 – Bubble Algae

1 – Surgeonfishes – the genus *Zebrasoma,* such as *desjardini, flavescens, scopas, veliferum, desjarini, xanthurus.*

1 & 4 – *Ecsenius bicolor* – The Bi-color Blenny (first trim algae to a short height).

1, 2 & 4 – *Ophioblennius atlanticus* – The Atlantic Red-lip Blenny.

2 & 5 – Surgeonfishes – the genus *Acanthurus, Ctenochaetus, Naso, Paracanthurus.*

6 – Surgeonfishes – the genus *Naso* and *Zebrasoma, e.g., xanthurum.*

1 & 6 – Rabbitfish, e.g., *Lo vulpinus* – The Foxface Rabbitfish.

1 & 2 – *Centropyge* angels – however can nip invertebrates.

1 – Gobies – *Amblygobius rainfordi* and *A. phalaena.*

1 & 6 – Emerald Crab *(Mithtax sculptus)* – can get big enough to be a danger to small fish.

1 – Banded-leg Hermit Crab *(Calcinus seurati).*

1 & 4 – Blue-legged Hermit Crabs *(Clibanarius tricolor)* – first trim hair algae to a short height.

2 – Red-legged Hermit Crab *(Calcinus tibicen).*

2 & 3 – Blue-spotted Hermit Crabs *(Clibanarius digueti)* – seem to prefer coralline to green algae.

2 & 3 – Polka-dotted Hermit Crab *(Phimochirus operculatus).*

1 & 2 – Scarlet-legged Hermit Crab *(Paguristes cadenati).*

2 & 3 – Sally Lightfoot crabs *(Percnon gibbesi).*

1 & 2 – Queen Conch – very good green algae eaters. Captive bred and very hardy.

1 – The Florida Sea Slug *Tridacia crispata* – the Lettuce Slug is very fond of Bryopsis. Also the slugs *Elysia ornata* and *Elysia diomedea.*

1, 2, 3 & 6 – Urchins are quite useful, however they bulldoze corals and strip coralline. The Black Longspine urchin, *(Diadema sp.,)* is one of the better choices.

1, 2 & 5 – *Astraea tectum* and *A. caelata* snails appear to be much better suited for longevity in closed systems than the *Turban* (Turbos) or *Nerita* species. Even though they are smaller and consume less than Turbos, they are a better choice. Quantity needed would be in the range of one per ten gallons in well-maintained systems.

4 & 5 – *Cerithium litteratum* (Snail).

1 – *Trochus niloticus* (Snail) – however they get too large for small aquariums. Very hardy.

Bear in mind large populations of copepods and amphipods can control many forms of minor green algae growth. These crustaceans proliferate if the system is somewhat devoid of carnivorous animals.

Check List

There are numerous reasons why unwanted forms of algae and cyanobacteria show up in the aquarium. I'll first attempt to briefly list all possible causes, some of which may not be applicable in some circumstances. The point here is to surface as many possibilities in as concise a manner as possible. Then let the reader pick those that may apply and then look through "Remedies" to find what may help resolved the problem. Some of the "Causes" noted below may seem extremely simple because they are. However, awareness of these causes goes a long way in preventing them in the first place.

Causes

- Less than adequate biological and chemical filtration methods used to set-up the aquarium.
- Areas of anoxic denitrification, such as in sandbeds, have been changed to anaerobic denitrification. These possibly hardened sandbeds, from misuse of calcium and buffer additives, now add plant nutrients instead of removing them.
- *Derbesia* and *Bryopsis plumosa* is caused by an over abundance of nutrients, e.g., ammonia, nitrate and phosphate, "and" incorrect spectrum. Bulk water nutrients or those on the surface of rock are perfect initial growth areas.
- Overfeeding, especially with products like flake or frozen foods can easily lead to "pockets" of uneaten food or partially metabolized waste products. These pocket areas are perfect breeding areas for cyanobacteria. Once started, sheet-like cyanobacteria can quickly spread, even in dimly lit areas.
- Brown diatoms are caused by silicate usually introduced into the aquarium via make-up water, the salt mix itself, silica sand, or something in the aquarium that may be composed of a silica substance.
- Dinoflagellates, free-swimming brown algae, can be introduced with live rock or coral. Under the right circumstances, i.e., water rich in organic material, they can multiply rapidly and form stringy globs of bubble filled goo. They can produce toxic side chemicals that further reduce water quality, making it more to the liking of other forms of unwanted algae.

- Overcrowding – the more fish the more food needs to be fed. This generates more waste products, resulting in a copious amount of bulk water nutrients. Some of which coat the rock in the system, establishing perfect sites for initializing new algae growth.
- It is quite normal for a new system to develop some forms of algae while the system is developing its initial microbial processes. Some may dissipate quite quickly, however others may gain a foothold. If the system has more nitrogen compound producing processes than efficient denitrification processes, it contributes to ongoing conditions favoring algae growth.
- The use of Diatomaceous Earth (DE), i.e., the skeletons of tiny single cell diatoms composed of silica, for filtering purposes can lead to brown algae problems.
- The use of unprocessed tap water, if containing plant nutrients, can lead to unwanted forms of algae.
- Medications can greatly reduce the efficiency of biological filtration. They destroy or inhibit bacteria and part of the result can be a rise in ammonia level. If experiencing any levels of ammonia, the safest thing to do is not add any further bio-load to the system until the source of the ammonia is resolved.
- The over use of certain additives can lead to unwanted algae growths. Not only do some additives directly influence algae growth, they may also indirectly affect them by increasing the level of seawater constituents above NSW levels, thereby negatively influencing the spectrum of the delivered light.
- Some salt mix formulas may contain algae nutrients.
- Lamps may be too old or incorrect spectrum.
- Failure to properly use protein skimming or activated carbon.
- Poor water movement.
- Crashing macroalgae growths quickly add to the nutrient level. The decomposition of dead algae adds nutrients and vitamins to the bulk water.
- Carnivorous fishes are eating too much of the microfauna, i.e., copepods and amphipods, that naturally eat some microalgae.
- Sand sifters/snails are dying off in hidden areas, not being removed, thereby adding to the nutrient level.
- Lack of water changes.
- Die-off of organisms living on or in newly installed live rock can result in an excessive initial nutrient level which in turn leads to giving algae a chance to gain a foothold.
- Phosphate can stick to any calcium carbonate material (live or dead rock/ substrate material), thereby providing fertile areas to get algae started.
- Starting systems with dead rock, e.g., dry tuffa/lava/concrete rock, and dry, dead sand. Dead rock has no beneficial bacteria living either in or on it. That may allow unwanted nutrient levels in the aquarium to rise without being processed by beneficial bacteria that would have normally inhabited live rock.
- Not keeping the mechanical filter clean.

- Excessive or unwanted forms of algae/cyanobacteria in refugia are generally caused by the type of lamp used for its lighting. Refugia are sometimes lit oppositely that of the main show aquarium. This encourages algae growth and helps stabilize pH during a timeframe when the carbonate buffering system is normally declining. Normal household fluorescent lamps are an excellent choice for this purpose because they are normally high in red spectrum. Therefore growth should be expected in such areas that combine such spectrum and nutrients.
- Molybdenum is an algae-enhancing trace element.
- Brown algae growth can be enhanced by the over usage of potassium iodide.
- Undergravel filter substrate may be too deep and water flow through it too slow. Gravel then becomes a dirt trap. This adds greatly to bulk water nutrient content.
- Not keeping a clean aquarium will add unwanted nutrients. Detritus breaks down into phosphate and nitrates very quickly.
- The green algae *Ostreobium,* sometimes found in coral skeletons, is nourished by excessive nitrate and phosphate levels.
- Cyanobacteria is caused by the accumulation of nutrients, mainly phosphate and nitrogen-based products such as ammonia. Both ammonia and phosphate are found in ample supply at the substrate-bulk water interface. Overfeeding, over-crowding, poor water movement, poorly balanced filtration, and lamps high in the red spectrum contribute greatly to its existence.
- Over usage of additives containing iron.
- Using a poor grade of calcareous material in a calcium reactor, such as a mix high in shell debris which normally contain high amounts of phosphate and heavy metals.

Remedies

Unwanted forms of algae are clearly a widespread problem. Herbivores, a more natural approach to algae control were discussed above and their use will not be repeated here. However, if using herbivorous animals, use commonsense stocking levels.

Over the past few decades I've heard of many remedies that have worked well for aquarists. Some may be extremely helpful in one situation and not as helpful in others. To be more concise, I've divided this section into subsections beginning with those remedies that are wide ranging and encompass hair algae problems. Then noted those that are more applicable to diatoms, dinoflagellates, and cyanobacteria.

General Controls/Comments

- Do not add trace elements of any kind until an algae problem is resolved.
- Check the label of any aquarium additive to see if it contains plant nutrients. If not sure, call or write the manufacturer.
- Adding some macroalgae, e.g., Caulerpa, may help assimilate some of the very nutrients pest algae utilize.

- Siphon off accumulating detritus.
- Test tap water before use. If needed, use RO or DI filtration prior to use.
- Vacuum up the upper most sand particles in living sand systems every three or four months. This will export any sand particles coated with calcium phosphate, a result of the mineralization process in the upper sand level.
- Medications to kill unwanted algae/cyanobacteria usually result in temporary results and a waste of money.
- If the local water company adds chloramine to your tap water, it should be removed prior to use, as it is a source of ammonia. The chloramine compound, a combination of ammonia and chlorine, is much more difficult to remove than simple chlorine. If unsure whether chloramine is a component of the tap water, call the local water supply company for information.
- Turf algae are often easily removed with tweezers. In areas where there are clumps, simply grabbing the center area with tweezers and applying a slight upward pulling motion removes a large clump.
- Use a turkey baster to blast clean turf algae areas. Keeping them free of detritus lessens nutrient accumulation in the bulk water.
- The use of a turkey baster to keep rock crevices clean of detritus will help lessen bubble algae (so will excellent water movement/surges of water).
- In UGF systems the correct water flow through the substrate should be approximately 6 to 8 times the volume of the aquarium. Also, having no more than two inches of gravel, gently stirring it once a week and vacuuming it every two weeks goes a long way in correctly maintaining a more algae-free system.
- Keep in mind that as hermit crabs grow, they need larger shells. Provide some empty shells in the system.
- *Bryopsis plumosa* is fairly easy to control because it is usually a single stem plant. Simply remove the entire plant structure at its base.
- Bubble algae cells or bubbles are held to the rock by tiny root-like appendages. A pointed, yet dull knife blade slipped under the cell(s) can pop them loose and the entire cell or group of cells can easily be removed from the aquarium unbroken. Try not to break them as the bubble contains many new little bubbles that can spread the problem.
- Areas of anoxic denitrification, such as in aragonite sandbeds, can be clogged by improperly maintaining the carbonate buffering system. Once precipitating carbonates clog these zones, no oxygen can penetrate, thereby reducing the area to an anaerobic condition. Sandbeds should occasionally be surveyed for slab-like conditions and if necessary, broken up into individual grains. Keep in mind anaerobic areas foster the denitrification process called Ammonification, which add plant nutrients in the form of ammonia compounds.

- Add a quality protein skimmer. A protein skimmer helps remove unwanted compounds prior to them breaking down. Also, the skimmers air borne particles, which tend to accumulate on the inside of the top cover of the unit, contain orthophosphate. Frequent cleaning of this area is highly recommended.
- Use activated carbon that is termed "phosphate free" or that test results indicate no leaching of phosphate. If not sure, soak and aerate a small portion of the activated carbon in either RO/DI or distilled water for a few hours and then test its water.
- Consider the use of ozone to help reduce nutrients and the transportation of alga cells.
- Molybdenum is an algae enhancer. Greatly reduce or discontinue its use. The same is true of iron.
- Vacuum substrate monthly or more often if needed.
- There are very specific types of equipment that accomplish the denitrification process, i.e., removal of nitrate directly from aquarium water. Most are small biological systems that use special additives to promote the process. Others may use small electrical currents or oxygen starvation methods. They are somewhat beneficial, yet maintenance intensive. Anything that fosters anaerobic denitrification is very risky to use in reef aquariums.
- Replace or wash mechanical filter media every few days.
- By maintaining very low levels of nitrate and phosphate, especially phosphate, the green algae *Ostreobium* can be restrained/controlled.
- If an unexpected algae bloom quickly occurs, check with the local water supplier to see if an anti-rust inhibitor was added before you consider other ways to resolve it. If so, treat with activated carbon and ion exchange molecular absorption filters.
- For a bad hair algae problem and where the rock is worth saving, I recommend using the following mechanical method. Depending on just how severe the problem is, remove any rocks that can be easily removed and brush them clean in a pail of water the same temperature and salinity as the aquarium. Temporarily store them in a container filled with some freshly prepared seawater. For the more difficult areas, wrap a long, wire handle, fairly stiff pipe/tube brush (something that would fit inside a half-inch ID tube) around a siphon hose. Make sure the brush portion is even with the end of the siphon hose. With the siphon action started, scrub the areas of hair algae. Pinch the end of the hose as needed to prevent too much water being drawn in when not necessary. Where hair algae gets caught on the brush, remove it by hand. Where it is drawn into the siphon tube, allow it to simply flow into the wastewater collection pail. When ready to replace the removed rock, scrub it with some white vinegar, rinse with seawater and place back in the aquarium. The vinegar is used to lessen any calcium phosphate that originally adhered to the rock surface. All water removed should be replaced with freshly prepared seawater. This labor-intensive method is usually quite successful. However, since phosphate is probably coating some of the rock, the algae will grow back. This method may have to be repeated a few times. Each time the hair algae will become less. It is a tough way to beat the

problem, but if the rock is worth saving this method works. Keep in mind the source of the excessive phosphate/nitrate needs to be eliminated or the problem will simply continue.

- Poor grades of salt mixes can initiate unwanted algae. Check not only their phosphate, silicates, and nitrate level, but also their ammonia level when freshly prepared. Do not be fooled by fancy claims. Contact long-term aquarists or aquarium societies for their comments and recommendations.
- Eliminate the desire to feed too often. If feeding more than once a day the chances are the fishes are not fully metabolizing their food intake before they again eat more food. Their waste may simply add undigested food/more waste products to the nutrient problem. Overfeeding, especially with liquid or frozen foods where much waste is incurred, is a sure-fire way of increasing aquarium nutrient level. Keep in mind 90% or more of liquid invertebrate food goes to waste.
- Excessive use of protein-based foods can quickly add up to a nitrate problem. Pay special attention to animal nutritional needs and feed a variety of foodstuffs.
- Use ion exchange, molecular absorption filter pads.
- There are forms of what is termed nitrate-removing media. Denitrifying bacteria colonize this porous material and when placed where aquarium water can flow slowly past it, a diffusion process will slowly over time remove nitrate. However, the process is more anaerobic than anoxic. Therefore, even though nitrate is reduced, the processes may be adding further nitrogen-based products, such as ammonia to the bulk water.
- Calcium reactors should always use a medium that is termed "phosphate" free or at least especially formulated for their use. There are a couple of excellent choices on the current market and you can contact me for further information.
- Aquascaping the aquarium correctly where there are sufficient spaces between rocks for good water movement will help prevent pockets of detritus from collecting.
- Utilize phosphate/silicate removing media and replace as required.
- Gorgonian black-band disease can be stopped by siphoning off the algae, brushing off remaining, and if possible, a 30-second freshwater dip.
- Consider a remotely connected plenum system/refugium.
- Some aquarists report that by simply placing an airstone in the aquarium it will lessen algae growth. Note that it takes a while before reductions in growth occur. Carbon dioxide reductions do work for some aquarists.
- Reconsider the desire to add "one more pretty fish."
- Some aquarists believe that mangroves will quickly absorb nutrients if used in their system. The truth is, nutrient removal is quite limited. They also require additional strong lighting to survive. If the system is already high in nutrients, additional lighting can increase further unwanted algae growth.

Diatoms

- When cleaning brown algae from the aquarium side panels, don't just wipe the viewing surfaces clean by using back-and-forth motion with a cleaning pad. This only distributes the diatoms/silica back into the bulk water where it's used again to produce more diatoms on the surface just cleaned. Instead, place the cleaning pad at the bottom of the surface to be cleaned and slowly slide the pad to the water's surface. Then quickly remove the pad and rinse in some clean water. Repeat as necessary. This way, the majority of the diatoms on the viewing surfaces will be "removed" from the aquarium. As for diatoms on a sand surface, siphon them out if they become mat-like and replace the water removed with freshly prepared seawater.
- Tap water, even if processed through an RO system, may still contain silicate. Very few RO units are effective at removing silicate. Even quality RO units remove only about 90% of incoming silicate. To remove all silicate from tap water, one must utilize the proper combinations of resins in a well-engineered DI unit. Salt mixes should also be tested. Remember, anything that may be composed of silica, e.g., silica sand, Diatomaceous Earth, or decorative rock has a slight solvency at the pH normally found in marine aquariums.

Dinoflagellates

- This free-swimming algae, can be reduced/eliminated with the use of ozone or a UV sterilizer. Also, flowing all aquarium water through a sponge filter, which should be cleaned twice a week, is helpful.

Cyanobacteria

- Increase water flow to remove dead spots/low current areas. Use a wavemaker to generate surge, which can prevent the settling of detritus in normally low flow areas, thereby eliminating concentrations of organic matter.
- Increase or provide efficient protein skimming to help remove the organic nutrients this pest needs to get started.
- Vacuum any/all existing cyanobacteria mats quickly and as often as necessary.
- Cyanobacteria is photosynthetic and prefers the red spectrum. Some aquarists report that after a few days of using only actinic light their cyanobacteria bloom had greatly diminished.
- The antibiotic Erythromycin sulfate or Tetracycline will kill cyanobacteria. Even though these chemicals may eliminate this unwanted pest the first time around, further problem blooms may become more resistant to the drugs. It is really much wiser to get to the root of the problem than to put antibiotics into the aquarium. Bear in mind that antibiotics negatively affect system biological filtration.
- Vacuum up the affected area's upper quarter inch of sand and replace with new sand. Often this removes an area where there was a concentration of organic material.

The "Almost" Algae-Free Aquarium

Sooner or later improperly established or maintained aquariums become a fertile environment for unwanted forms of algae and cyanobacteria. It's a far too common happening and one that can be avoided with some forethought – a goal of this booklet. By grasping what conditions might bring on devastating battles with unwanted algae before they actually occur, most of these battles could be avoided. Of course, this is not to say many new aquariums will not exhibit some form of minor alga growth while microbial processes are establishing themselves. However the key word there is "minor."

Let's be straightforward – there's no such thing as an algae-free aquarium. Nor is there an algae-free reef in the wild. Yet it is possible to have a closed system that remains quite free of unwanted forms of algae "if" attention is paid to several fundamental factors, e.g., biological filtration, chemical filtration, spectrum and intensity, and bio-load.

It should now be quite clear the initial cause of unwanted algae is excessive amounts of the basic nutrients, "and" the impact of light. Once started, these unwanted forms of algae become labor intensive and costly to overcome. Therefore, it makes more sense to prevent them from gaining a foothold than engaging in all-out war to overcome them once established. The first and foremost important husbandry factor is construction of a system that contains a truly beneficial balance of microbial processes.

Biological Filtration

During the aquarium's first few months of performance, different microbial colonies having different functions are forming. While establishing their own equilibrium processes and linking with other microbial colonies, some nutrients may become sufficient enough to generate minor outbreaks of algae. These outbreaks usually have short life spans if biological filtration becomes balanced in favor of using incoming nutrients for energy, rather than accumulating and becoming stored in the sandbed/rock or bulk water.

The most probable cause of sustained plant growth in closed systems is the lack of balanced microbial processes. In such cases, mineralization and nitrification found in oxic areas and inefficient denitrification processes found in anaerobic areas outweigh (volume-wise) the efficient denitrification processes found in anoxic areas. Such a skewed microbial system then favors nitrogen storage, usually seen as unwanted plant growth. Systems that use trickle filters, fluidbed filters, incorrectly established sandbeds, and those overcrowded with live rock, discharge more nitrogen-laden products than what can be naturally and efficiently utilized. The result is accumulation of nitrogen compounds seen as algae growth.

Bear in mind that the end product of trickle and fluidbed filters is nitrate, since nitrification is their primary purpose. Deep sandbeds directly on the aquarium bottom favor nitrogen assimilation (plant growth) because the bed contains more anaerobic than anoxic volume. Shallow and/or fine grain sandbeds usually lack sufficient anoxic area to adequately reduce all incoming nutrients. And, the more water currents and fauna that disturb them, the more oxic they become thereby rendering less microbial equilibrium/balance.

Also consider the fact that live rock contain more anaerobic area than oxic and anoxic areas. Their oxic area, where nitrification occurs, consists of a surface area and probably a very shallow oxic zone consisting of a few millimeters in depth. Of course this depends upon the density of the rock. The more dense the material, the less the oxygen gradient. Below the oxic zone, again depending on its density, there is usually an extremely shallow anoxic zone where very efficient denitrification occurs. Below this zone the majority of the material becomes anaerobic. And of course, anaerobic areas are to be avoided if at all possible because that is where the least efficient form of denitrification occurs. Therefore, the majority of rock material is actually skewed towards producing nitrogen-laden products!

When there is a copious amount of areas for mineralization and nitrification, there is an ample supply of nitrate. Nitrate is reduced in anaerobic areas by a denitrification process called Ammonification. This simply results in further nitrogen-laden products, (i.e., ammonia). In many improperly microbial balanced systems it's possible to have small nitrate levels and still have unwanted plant growth because nitrogen from the widespread Ammonification process is now being used for initiating/plant growth. In systems designed to have ample anoxic areas, such as in plenum systems, the nitrate ion is reduced to its basic components via the "Destructive" form of denitrification. In addition, phosphorus is used more efficiently resulting in less production of phosphate. When less phosphate is produced, there will also be less precipitation of it in the form of calcium phosphate, which often occurs when aquarists misjudge the application of calcium and buffer additives.

If aquarists first construct a truly efficient biological filtration system, they will have much less unwanted algae problems in both the short and long term. Unfortunately, too many aquarists still fail to realize that the nitrification cycle is only a small portion of the overall microbial processes in the aquarium and/or that there are two forms of denitrification. There's no doubt the nitrification cycle is essential in the aquarium so as to convert ammonia laden waste products of animals into less toxic compounds such as nitrate. Yet, if nitrate is not "efficiently" processed, algae will eventually assimilate the nitrogen compounds generated. This is the reason to foster anoxic areas where "Destructive" denitrification processes are prevalent, not Ammonification denitrification as in anaerobic areas.

Remember that as plant nutrients become more prevalent, algae cells brought into the aquarium on live rock and coral are awakened so to speak. They either assimilate nitrate and reduce it to ammonia for growth or uptake ammonia directly. When that occurs, more and more green becomes visible in the aquarium. Of course, if green is the favorite color, then who cares. But usually that's not the case.

Microbial equilibrium and balance is a common sense husbandry approach. How to accomplish this type system is clearly explained in my *"Live Sand Secrets"* booklet. It explains how a plenum system provides not only for mineralization and nitrification, but also supplies extensive anoxic areas where more efficient denitrification and use of phosphorus occurs.

The fact that a plenum sandbed can easily provide all the biological filtration necessary also reduces the need for live rock. Limiting the use of live rock and/or possibly supplementing its use with aquacultured rock makes further sense. By doing so, both Mother Nature and aquarists' pocketbooks benefit! Even if installing a plenum in the main system is not desirable, consider connecting an auxiliary plenum system to the main show aquarium.

In fact, my next system is on the drawing board. It will be a 180 gallon reef system with about one inch of sand on the bottom. Underneath and connected to this system, a 55-gallon aquarium will contain a plenum sandbed and nothing else but a small water pump. Protein skimming will be utilized via a separate, standalone unit. A calcium reactor will provide all but a few natural seawater constituents such as magnesium and iodide. No live rock will be purchased. Aquascaping will consist of different coral animals, all presently growing on existing branching rock in smaller aquariums. I expect the connected plenum system to adequately maintain a nutrient-poor bulk-water condition that will naturally keep algae at a minimum.

Without a doubt, a truly efficient biological system is the first CORRECT step to take in setting up an "almost" algae-free aquarium.

Chemical Filtration

Getting started with the best of the best biological filtration method can not be faulted. Nor should certain forms of chemical filtration be negatively debated. Yet, some aquarists question the value of protein skimming and use of activated carbon. I find that kind of thinking counterproductive to proper husbandry procedures. Actually, these chemical filtration methods go hand-in-hand with biological filtration, something like bread and butter!

I consider protein skimming and the use of activated carbon as the two most important chemical filtration methods. Dissolved organic carbon (DOC) compounds, nitrate, and inorganic phosphate tend to increase as the captive system ages even though not obvious in fairly new systems. All too often the aquarist fails to recognize an increasing biological load. An aquarium with a few fishes, fed once everyday, may seem stable for a long period of time. The aquarium water may appear clear, with little or no unwanted plant growth. Dissolved nutrients such as nitrate and phosphate may appear to be well below what the aquarist perceives as critical levels.

However, there's the inclination to add one more attractive specimen, overfeed, or allow maintenance to lag. Furthermore, bacteria are constantly being renewed, as their normal life cycle is quite short. Dead bacteria and those that take their place, continue to a certain degree, to add biomass to the aquarium even though the aquarist may not be adding new specimens. These factors effect any aquarium relying solely on biological filtration and the aquarium may become a breeding ground for unwanted algae growths. Not only do accumulating dissolved nutrients help give algae a start, they also skew the color of the light as it travels through water.

To efficiently accomplish the chemical filtration adsorption process, both protein skimming and activated carbon should be used. The reasons for this are clearly detailed in great length in my *"Protein Skimming and Activated Carbon Secrets"* booklet.

Nonetheless, protein skimming and activated carbon effectively remove substances that are detrimental to water quality in closed systems. Both should be considered as additional filtration methods or water polishing methods. They are not a substitute for proper biological filtration and continued water changes. When combined with efficient biological filtration they could be considered something like shock absorbers in a car. They smooth out those maintenance bumps that all systems incur sooner or later.

While on chemical filtration, many aquarists improperly dose calcium and buffers, or add some of this or that just because it seems like a good thing to do. They simply do not realize the negative impact it may have on light transmission. There's another practice that I don't agree with and that is precipitating phosphate out of solution with the use of Kalkwasser. I won't go into all the negative effects improper dosing of Kalkwasser can have on the carbonate buffering system, but using it to remove phosphate, in my opinion, only places it in areas that may benefit new algae growth.

It is my position that at whatever pH Kalkwasser precipitates a phosphate-coated ion, whether that is calcite or aragonite, as soon as system pH falls below the pH at which it was created, it will start to dissolve. That simply places a readily useable amount of phosphate in position to be used by future algae growth every time the system's lights go off/pH drops. Only in systems where there is diligent control and proper ongoing dosage of Kalkwasser is phosphate controlled for the long term. Unfortunately, that's not the case in many systems.

Spectrum & Intensity

A third area of concern is lighting. The color of light reaching invertebrate and plants is important. Corals like the blue-green spectrum. Plants like the red spectrum. Selecting a sufficient wattage and spectrum for lamps is a good beginning. Yet all the proper spectrum and intensity is of no value if the medium it has to travel through, (i.e., the bulk water), is skewed with nutrients and/or constituents that exceed those found in natural seawater.

For instance, excessive levels of nutrients and calcium bend/refract the light waves passing through water. What may have been a blue-green wavelength to begin with, can easily be shifted to more red band wavelength by the time it reaches the target organism. To further complicate matters, many lamps that originally produced an excellent spectrum when new begin to lose their ability to do so when six months to a year old. Depending upon the type of lamp/ brand, it's better to install new lamps when recommended, than to wait a few additional months.

Bio-load

Aquarists should make a decision as to what species and quantity will ultimately live in the system before setting it up. They must keep in mind what the system can adequately support in the way of bio-load. If an aquarist attempts to stuff six pounds of waste into a five pound bag, he or she will have to resort to high tech equipment to process the excess! In the first edition of *"Live Sand Secrets"* I stressed a very conservative rule-of-thumb as to how many fishes can be properly maintained in a captive natural system. I have always believed low to moderate bio-loads provided for long-term success. This very conservative belief noted that fishes represent the majority of bio-load in reef and fish-only aquariums. It recommended not exceeding one inch (2.5 cm) of fish, not counting the tail, per ten gallons of water in reef aquariums. In the fish-only aquarium it recommended one inch of fish per five gallons of water.

However that has recently changed. Sam Gamble has been trying to convince me over the past couple of years that plenum-equipped sandbed systems can carry a much greater bio-load than a rock-based system with a sandbed directly on the aquarium bottom. After reviewing all the data and feedback from aquarists with more heavily stocked plenum systems, I now agree. The reason for this increased capacity is the plenum system's far greater volume of anoxic zones. In fact, it's much more than what is found in sandbeds of any depth directly on an aquarium bottom or that inside of live rock. This greater anoxic area allows highly efficient bacteria that prefer this zone to occur in greater numbers. Therefore, most or all of the incoming nutrients are efficiently used/processed.

Yet, if the desire to have a "very" large bio-load and/or general husbandry procedures decline, don't blame algae problems on the plenum method! It doesn't guarantee algae-free systems. However, it does provide what no other so-called natural closed system can provide – a more efficient balance of microbial processes. In fact, it's even more efficient than what is found in Mother Nature!

Closing

In retrospect, if what was described in my first two booklets were utilized to set up all aquariums, this third booklet wouldn't sell many copies! However there will always be some aquarists not adequately updated.

In my first booklet it clarified what I believe to be the best form of biological filtration. The second booklet explored what I consider the best forms of chemical filtration. By building upon the information disseminated in those booklets an aquarist could establish a very efficient microbial foundation that would be able to greatly limit unwanted forms of algae or cyanobacteria, both in the short and long-term. Bear in mind that the foundation of any home or aquarium is extremely important to its long-term existence. Give it the attention it deserves!

As complete as I hope the text in this booklet is and that I accomplished my "goals," if you have any questions or additions you'd like to see addressed in a future edition, contact me through any of the publications I write for or the INTERNET at saltcorner@compuserve.com.

I should also note that after six years of research, Sam Gamble and I have just completed writing a leading edge book that will take its reader into the next millennium of aquarium husbandry. *The New Wave* will also dispel many of the myths that have besieged the aquarium industry for many decades and denote new, and astonishing processes that hold much promise for the well-being of closed systems. Contact me for further information.

Before I close, I want to thank Bob Fenner for his input on the taxonomy of plants and suggestions for further reading. I truly believe we both have the same common goals – to better the environment in the wild and in closed systems.

Bob Goemans, Ph.D.

Bibliography/Further Reading

Baugh, Thomas M. 1988. *Caulerpa prolifera;* an attractive species which does well in marine aquaria. FAMA 5/88.

Blackburn, Wayne. 1989. Plants in the marine system. FAMA 7/89.

Bold, Harold C. & Michael J. Wynne. Introduction to the Algae. Prentice-Hall, NJ.

Brelig, Allen. 1993. Plants in the reef system. FAMA 6/93.

Dawson, E. Yale. 1966. Marine Botany; An Introduction. Holt, Rinehart, Winston. NY.

Debelius, Helmut & Hans A. Baensch. 1994. Marine Atlas, Mergus, Germany, pp. 214 - 324.

Delbeek, J.C. & J. Sprung. 1994. The Reef Aquarium. Ricordea Publ., Coconut Grove, FLA. V1, pp 257 - 281.

Dewey, Don. 1979. Marine algae; a key ingredient to the successful saltwater system. FAMA 8, 9/79.

Emmens, C.W. 1991. Algae. FAMA 8/91.

Fenner, Bob. 1989. Some notes on algae and their control. FAMA 6/89.

Fenner, Bob. 1990. Macroalgae for marine aquaria. Pet Dealer 2/90.

Fenner, Bob. 1997. The Conscientious Marine Aquarist. Microcosm, VT, pp. 133 - 139.

Gamble, Sam. 1995. Algae curse: a new view. FAMA 8/95.

Giovanetti, Thomas A. 1989. *Caulerpa*; enemy of the miniature reef aquarium? FAMA 10/89.

Glodek, Garrett. 1993. A lesson in plant taxonomy and systematics. FAMA 11/93.

Glodek, Garrett. 1996. All you ever wanted to know about marine plants (selected bibliography). FAMA 1/96.

Goemans, Bob. 1993. Natural Nitrate Reduction (NNR). V8, Part 1 - 4, Marine Fish Monthly, issues 2 - 5.

Graff, Rick. 1993. Getting control of microalgae; how to deal with this problem in reef tanks. AFM 1/93.

Hoff, Frank H. Marine algae of the genus *Caulerpa*. FAMA 10/83 & 4/84.

Holliday, Les. 1999. Water in the Desert. Practical Fishkeeping. Feb. pp. 48 - 51.

Hunt, Philip. 1996. Marine algae: friends and foes. Parts 1 & 2. TFH 9, 10/96.

Littler, Diane Scullion & Mark M., Bucher, Katina E. & Norris, James N. 1989. Marine Plants of the Caribbean, A Field Guide From Florida To Brazil. Airlife Publishing, England.

McGregor, Glenn. 1996. Cyanobacteria. FAMA 8/96.

Michael, Scott. 1995. All that algae; If it makes you feel any better, this is a common problem. AFM 8/95.

Paletta, Michael. 1990. Eliminating problem algae. SeaScope V. 7 Fall.

Prasek, Ed. 1995. Mirror & slime algae; are they driving you crazy too? TFH 7/95.

Smit, George. 1987. The ecological marine aquarium, part four: The use and benefits of live rock and *Caulerpa* in marine aquariums. FAMA 8/87.

Sprung, Julian. 1989. Reef Notes (on reproductive loss of *Caulerpa*). FAMA 2/89.

Tullock, John H. 1983. Growing marine "plants." FAMA 3/83.

Wecsler, Lawrence H. 1975. Algae in the marine aquarium. TFH 2/75.

Wilkens, Peter. 1992. Green water in the marine aquarium; the sexual reproduction of *Caulerpa* algae. TFH 2/92.

Wilkens, Peter. 1994. Red calcareous algae: Creator of coral reefs. Aquarium Frontiers Spr. 94.